MECHANICS MEMORY JOGGER

A-LEVEL APPLIED MATHEMATICS
EDEXCEL

A LEVEL POCKET GUIDES

THE MATHS CLINIC
Copyright © 2022 The Maths Clinic
All rights reserved.
ISBN: 979-8813774225

DEDICATED
TO THE SERIOUS STUDENT

"The building blocks of understanding are memorisation and repetition"

-Barbara Oakley-

CONTENTS

THE Q&A WAY	7
QUESTIONS - MODELLING	8
QUESTIONS - CONSTANT ACCELERATION	8
QUESTIONS – FORCES AND MOTION	9
QUESTIONS – VARIABLE ACCELERATION	10
QUESTIONS –MOMENTS	11
QUESTIONS – FORCES AND FRICTION	11
QUESTIONS – PROJECTILES	12
QUESTIONS – APPLICATIONS OF FORCES	13
QUESTIONS – VECTORS IN KINEMATICS	14
ANSWERS - MODELLING	15
ANSWERS - CONSTANT ACCELERATION	16
ANSWERS – FORCES AND MOTION	18
ANSWERS – VARIABLE ACCELERATION	23
ANSWERS –MOMENTS	25
ANSWERS – FORCES AND FRICTION	27
ANSWERS – PROJECTILES	30
ANSWERS – APPLICATIONS OF FORCES	33
ANSWERS – VECTORS IN KINEMATICS	36

THE Q&A WAY

Drawing from years of teaching Mathematics, The Maths Clinic has devised a foolproof method to master mathematical concepts for success in exams.

Armed with pen and notepad, follow the Q&A way as you revise Mechanics for the A Level Applied Mathematics examination. It is highly recommended that you spend a day copying out the facts at least twice in order to familiarize yourself with this knowledge.

The book is divided into two sections – Questions and Answers and covers the syllabus for Paper 3 of A Level Mathematics. Topics included here are Modelling, Constant Acceleration, Forces and Motion, Variable Acceleration, Moments, Forces and Friction, Projectiles, Application of Forces and Vectors in Kinematics.

The **Question section** breaks up each topic into a selection of questions on key facts, formulae, working steps and derivations.

The **Answer section** answers these questions listing the key facts, formulae, working steps and derivations.

In the hours before your exam, it would serve you well to quiz yourself by working through the Question section a few times.

.

MECHANICS MEMORY JOGGER 8

QUESTIONS - MODELLING

MODELLING ASSUMPTIONS
1. Describe a particle.
2. Describe a uniform body.
3. Describe a light object.
4. Describe a light, inextensible string.
5. Describe a smooth surface.
6. Describe a rough surface.
7. Describe a thin wire or rod.
8. Describe a smooth, light pulley.

QUESTIONS - CONSTANT ACCELERATION

MOTION GRAPHS
Distance – Time graphs
1. Describe how a distance time graph is plotted.
2. What is the gradient?
3. What does straight line graph indicate?
4. What does a curve indicate?

Velocity -Time graphs
1. Describe how a velocity-time graph is plotted?
2. What is the gradient?
3. What does straight line graph indicate?
4. What is the area under the graph?
5. What does negative area mean?

CONSTANT ACCELERATION
1. State the 5 formulae for motion in a straight line with constant acceleration.
2. State the units for distance, time, velocity and acceleration.

VERTICAL MOTION UNDER GRAVITY
1. State the acceleration due to gravity g, for a falling body and a particle projected vertically upwards.
2. What is the recommended degree of accuracy of the answer when using g in a problem?

QUESTIONS – FORCES AND MOTION

FORCES AND MOTION
1. State the weight of a body W given its mass m. Indicate the forces acting on the body.
2. Define the normal reaction.
3. State the equation of motion.
4. What is the unit of force?
5. Define thrust.
6. Define tension.
7. Indicate the forces acting on a body in motion.
8. For a car pulling a trailer with a tow bar, indicate the tension T in the tow bar when the car accelerates.
9. For a car pulling a trailer with a tow bar, indicate the tension T in the tow bar when the car brakes.
10. For a car pulling a trailer with a tow bar, indicate the tension T in the tow bar if the tow bar breaks.

FORCES AS VECTORS
1. Define the magnitude of a force $F = ai + bj$.
2. What is a unit vector?
3. How do you find the resultant of forces given in vectors.
4. What is the resultant of forces given in vectors if a particle is in equilibrium?
5. Given that force is a vector for a moving particle, describe acceleration.

PULLEYS
1. Indicate the tension T when two particles are connected by a light, inextensible string which passes over a smooth light fixed pulley.

MECHANICS MEMORY JOGGER

2. Write the two equations of motion when the particles are released.

CONNECTED PARTICLES
1. Indicate the tension T when two particles are connected by a light, inextensible string which passes over a smooth light fixed pulley - where one particle is on a rough horizontal table and the other is suspended from the table.
2. Draw a force diagram for the arrangement and write the two equations of motion when the particles are released.
3. What is the force exerted by the string on the pulley?
4. Indicate how you would find the force exerted by one object on another when 2 objects are stacked on each other.
5. Describe the force diagram for a lift and state the equation of motion to find the force R acting on the person of mass m kg by the lift as it is moving upwards with acceleration a.

QUESTIONS – VARIABLE ACCELERATION

VARIABLE ACCELERATION
1. What variables can be written as functions of time?
2. What is the derivative of the displacement?
3. What is the derivative of the velocity?
4. What is the integral of the acceleration? How is the integration constant found?
5. What is the integral of the velocity? How is the integration constant found?
6. How can you find the maxima or minima of a function of time?
7. What indicates that an object is instantaneously at rest?

DERIVATION OF SUVAT FORMULAE
1. If $v = \int a\, dt$ derive $v = u + at$
2. If $s = \int v\, dt$ derive $s = ut + \frac{1}{2}at^2$
3. Given that $v = u + at$ and $s = ut + \frac{1}{2}at^2$ derive $v^2 = u^2 + 2as$

4. Looking at the velocity time graph, derive the distance travelled $s = \frac{(u+v)t}{2}$

QUESTIONS –MOMENTS

MOMENTS
1. Define a moment of a force.
2. State the formula for the moment of a force.
3. What is the unit of a moment of a force?
4. Define equilibrium of a horizontal rod.
5. Describe the vertical resolution of forces when a uniform rod of weight W is hanging supported by rope or cables.
6. Describe the vertical resolution of forces when a uniform rod of weight W is resting on supports.

CENTRE OF MASS
1. What is the assumption made for the centre of mass of a uniform rod?
2. What is the assumption made for the centre of mass of a non-uniform rod?
3. Describe what happens when a rod is on the point of tilting?

QUESTIONS – FORCES AND FRICTION

RESOLVING FORCES
1. State the horizontal and vertical components of a force F acting at an angle θ to the horizontal.
2. Resolve the forces acting on an object on a rough plane inclined at an angle θ.

FRICTION
1. Define friction.
2. Define the coefficient of friction μ.
3. State the maximum frictional force between two surfaces given the coefficient of friction μ and the normal reaction R.
4. What is the direction of the frictional force?
5. Draw a force diagram showing the frictional force.

SLOPING PLANES
1. Draw a force diagram of an object moving on a rough plane inclined at an angle θ showing its weight, normal reaction and friction.
2. If the angle of inclination of the slope is given by $\tan\theta = \frac{4}{5}$, what are $\sin\theta$ and $\cos\theta$?
3. When does an object start to slip down a slope?

QUESTIONS – PROJECTILES

VERTICAL PROJECTION
1. For objects projected upwards, state the acceleration due to gravity.
2. Define the time of flight.
3. Define the speed of projection.
4. State the displacement of an object that is projected and returns to its original position.
5. State the displacement of an object if the original projected position is X m above the returning position.
6. State the displacement of an object if the time of flight is required above a height h.

HORIZONTAL PROJECTION
1. State the force acting on a particle projected horizontally.
2. State the horizontal and vertical components of a particle projected horizontally from a height h with a velocity $v\ ms^{-1}$ landing at a distance d.
3. What is the distance of a particle from its point of projection?

PROJECTION AT AN ANGLE
1. State the horizontal and vertical components of a particle projected at an angle θ to the horizontal with an initial velocity U.
2. Define the range of the projectile.
3. Define the time of flight of the projectile.
4. State the velocity when the projectile reaches its greatest height.

PROJECTILE MOTION FORMULAE
1. Derive the time of flight.
2. Derive the horizontal range.
3. Derive the greatest height reached.
4. Derive the equation of the trajectory.

QUESTIONS – APPLICATIONS OF FORCES

STATIC PARTICLES
1. When is a static particle said to be in equilibrium?
2. State the condition for a body on a rough surface said to be in limiting equilibrium.
3. State the condition for a body on a rough surface said to be at rest and in equilibrium.

STATIC RIGID BODIES
1. Draw the force diagram for a uniform ladder leaning against a smooth wall with one end resting on rough, horizontal ground. Resolve the forces vertically and horizontally.
2. Draw the force diagram for a uniform ladder leaning against a smooth wall with one end resting on rough, horizontal ground when the ladder is in limiting equilibrium. Resolve the forces vertically and horizontally and use moments to describe the situation.
3. Draw the force diagram for a uniform ladder leaning against a rough wall with one end resting on rough, horizontal ground when the ladder is in limiting equilibrium. Resolve the forces vertically and horizontally and use moments to describe the situation.

QUESTIONS – VECTORS IN KINEMATICS

VECTORS IN KINEMATICS
1. Define a position vector. What is its magnitude?
2. Define a velocity vector. What is its magnitude?
3. State the vector equation for a position vector at time t that starts from a point with position vector r_0 and moves with a constant velocity v.

VECTORS AND BEARINGS
1. Name the unit vectors for due east and due north.
2. State how you would show that two particles collide?

CALCULUS WITH VECTORS
1. Chart the relationship between displacement, velocity and acceleration to solve problems in 2D kinematics
2. What would the constant of integration be for vector integration?

ANSWERS - MODELLING

MODELLING ASSUMPTIONS
1. A **particle** is a small object of point mass which has no dimensions.
2. A **uniform body** is one that has its mass evenly distributed and concentrated at a single point called the centre of gravity.
3. A **light object** is an object considered to have zero mass.
4. A **light, inextensible string** is one that does not stretch and has zero mass. The tension in the string is constant and masses attached to the ends of the string will move with the same acceleration.
5. A **smooth surface** is one where there is no friction between the surface and an object moving over it.
6. A **rough surface** is one where there is friction between the surface and an object moving over it.
7. A **thin wire or rod** is considered to have zero mass.
8. A **smooth and light pulley** is considered to have zero mass and presents no friction to a string passing over it. The tension will be the same on both sides of the pulley.

ANSWERS - CONSTANT ACCELERATION

MOTION GRAPHS
Distance – Time graphs
1.

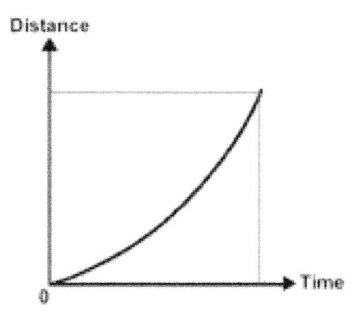

The motion of an object can be represented by a distance - time graph. Time is always plotted on the x–axis, distance is plotted on the y–axis.
2. The gradient of a distance - time graph is the velocity.
$$Speed = \frac{Distance}{Time}$$
3. A straight line indicates a constant speed.
4. A curve indicates an acceleration or a deceleration.

Velocity -Time graphs
1.

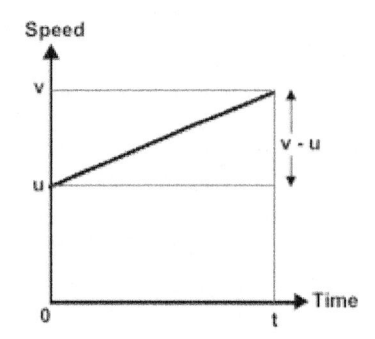

The motion of an object can be represented by a speed - time graph. Time is always plotted on the x–axis, speed is plotted on the y–axis.

ANSWERS

2. The gradient of a speed - time graph is the acceleration
   Acceleration $a = \frac{(v-u)}{t}$
3. A straight line indicates a constant or uniform acceleration.
4. The area under a speed – time graph is the distance travelled. Distance $s = \frac{(u+v)t}{2}$
5. Negative areas imply negative distance and signal a negative direction.

CONSTANT ACCELERATION

1. Five formulae for motion in a straight line with constant acceleration are:
   $v = u + at$
   $s = \frac{(u+v)t}{2}$
   $v^2 = u^2 + 2as$
   $s = ut + \frac{1}{2}at^2$
   $s = vt - \frac{1}{2}at^2$

2. The units for distance, time, velocity and acceleration are:
   s – displacement or distance (metres m)
   u – initial velocity (metres per second ms^{-1})
   v – final velocity (metres per second ms^{-1})
   a – constant acceleration (metres per second per second ms^{-2})
   t – time (seconds s)

VERTICAL MOTION UNDER GRAVITY

1. For falling objects, $g = +9.8 \ ms^{-2}$
   For objects projected upwards, $g = -9.8 \ ms^{-2}$.
2. The recommended degree of accuracy for the answer when using $g = +9.8 \ ms^{-2}$ in a problem is 2 significant figures.

MECHANICS MEMORY JOGGER

ANSWERS – FORCES AND MOTION

FORCES AND MOTION
1. The weight of an object W is the force due to gravity acting on it. $W = mg$. W is the force in Newtons, m is the mass of the object in kg and g is the acceleration due to gravity $9.8\ ms^{-2}$.

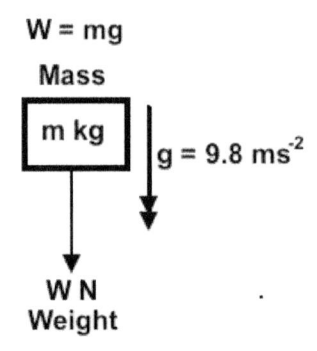

$$W = mg$$

2. The normal reaction R is the force equal in magnitude to the weight W of the object.
$$R = W$$

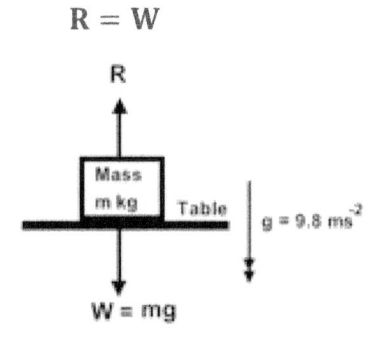

3. $F = ma$ is the equation of motion. F is the force in Newtons, m is the mass of the object in kg and a is the acceleration in ms^{-2}.
$$F = ma$$

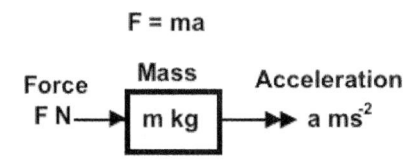

ANSWERS

4. The unit of force is a Newton.
5. Thrust is the force exerted on an object in order to push the object.
6. Tension is the force that is transmitted through a string when it is pulled tight by forces acting from opposite ends.
7. Indicate the forces acting on a body in motion.
   The resultant force is symbolically represented by R and an arrow indicating the direction of resolution:
   $R(\uparrow), R(\downarrow), R(\rightarrow), R(\leftarrow)$.

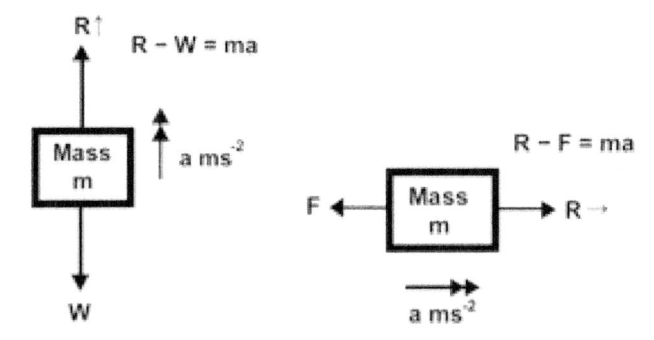

   $R - W = ma$

   $R - F = ma$

8. For a car pulling a trailer with a tow bar, indicate the tension T in the tow bar when the car accelerates.

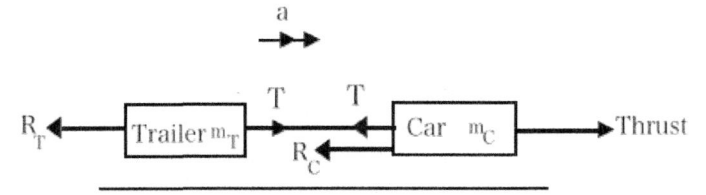

   Mass of Car and Trailer $= m_C, m_T$
   Resistances to motion of Car and Trailer are$= R_C, R_T$
   Tension in tow bar $= T$
   Acceleration $= a$
   Thrust $= F$
   Equation of motion of the car $R(\rightarrow),$:
   $F - T - R_C = m_C \times a$
   Equation of motion of the trailer $R(\rightarrow),$:
   $T - R_T = m_T \times a$

MECHANICS MEMORY JOGGER   20

9. When the car brakes, the acceleration changes direction and the tension in the tow bar works outwards from the tow bar.

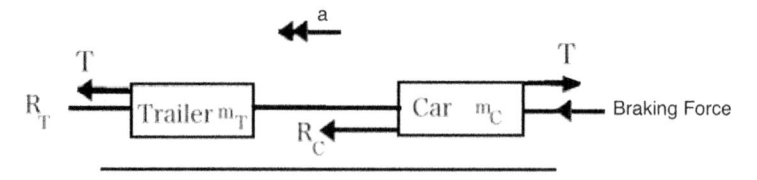

```
                        a
                       ←←
        T                              T
  R_T ←   ┌─────────┐         ┌───────┐
      ←───│Trailer m_T│         │ Car m_C│ ←── Braking Force
         └─────────┘  R_C ←    └───────┘
```

Braking force = F
Equation of motion of the car $R(\leftarrow)$.
$R_C + F - T = m_C \times a$
Equation of motion of the trailer $R(\leftarrow)$
$T + R_T = m_T \times a$

10. If the tow bar breaks, consider the motion of the car and trailer separately with tension T being 0.

FORCES AS VECTORS

1. If a force $F = ai + bj$, the magnitude of $F = |F| = \sqrt{a^2 + b^2}$
2. A unit vector of $F = ai + bj$ is given by $\frac{F}{|F|} = \frac{ai+bj}{\sqrt{a^2+b^2}}$.
3. The resultant is the sum of the forces. Given forces in vectors, add their i and j components for a resultant vector.
4. If a particle is in equilibrium, the resultant of forces is zero. add the i and j components of the forces and equate them to 0.
5. $F = ma$ where m, the mass of the object, is a scalar. If force is a vector then acceleration is also expressed as a vector $a = \frac{F}{m}$.

PULLEYS

1. When two particles are connected by a light, inextensible string which passes over a smooth light fixed pulley, the tension T in the string is equal on either side of the pulley and is directed towards the pulley.
2. Write the two equations of motion when the particles are released.

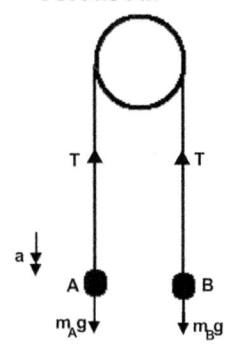

Acceleration $= a$
Weights A and B $= m_A g, m_B g$
$m_A > m_B$ so the direction of motion is downwards at A
Tension in the string $= T$
Equation of motion of particle A $R(\downarrow)$:
$m_A g - T = m_A \times a$
Equation of motion of particle B: $R(\uparrow)$
$T - m_B g = m_B \times a$

CONNECTED PARTICLES

1. When two particles are connected by a light, inextensible string which passes over a smooth light fixed pulley - where one particle is on a rough horizontal table and the other is suspended from the table, the tension T in the string is equal on either side of the pulley and is directed towards the pulley.

MECHANICS MEMORY JOGGER   22

2. Force diagram for the arrangement and the two equations of motion when the particles are released:

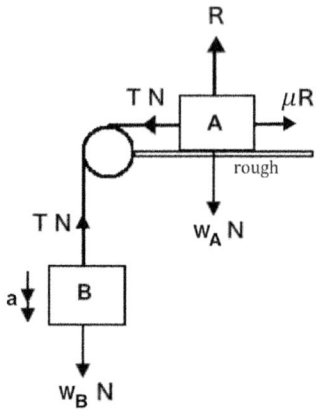

Acceleration $= a$
Weights A and B $= W_A, W_B$
Masses A and B $= m_A, m_B$
Motion is downwards at B and left at A
Tension in the string $= T$
Equation of motion of particle B $R(\downarrow)$:
$W_B - T = m_B \times a$
Equation of motion of particle A: $R(\leftarrow)$
$T - \mu R = m_A \times a$

3. The resultant force exerted by the string on the pulley acts at 45° and is $2T \cos 45$.

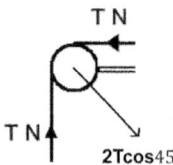

ANSWERS

4. In order to find the force exerted by object B on object A of mass m, examine the equation of motion of A.

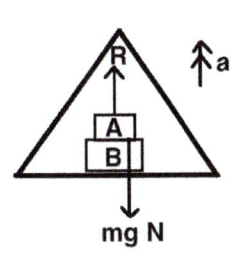

Equation of motion of particle A: $R(\uparrow)$
$R - mg = m \times a$

5. Describe the force diagram for a lift and state the equation of motion to find the force R acting on the person of mass m kg by the lift as it is moving upwards with acceleration a.

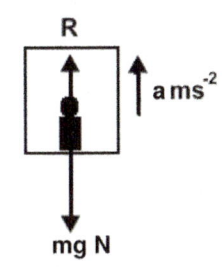

Equation of motion of person: $R(\uparrow)$
$R - mg = m \times a$

ANSWERS – VARIABLE ACCELERATION

VARIABLE ACCELERATION
1. Displacement, velocity and acceleration can be written as functions of time.
2. The derivative of the displacement is the velocity $v = \dfrac{ds}{dt}$.
3. The derivative of the velocity is the acceleration
$$a = \dfrac{dv}{dt} = \dfrac{d^2s}{dt^2}$$

4. The integral of the acceleration is the velocity $v = \int a\, dt$
   Use the initial velocity at time t to determine the integration constant.

5. The integral of the velocity is the displacement $s = \int v\, dt$
   Use the initial displacement at time t to determine the integration constant.

6. Use differentiation to find the maxima or minima of a function of time.
   To find the maxima or minima of $s = f(t)$, differentiate and solve $f'(t) = 0$.

7. An object is instantaneously at rest when its velocity $= 0$.

DERIVATION OF SUVAT FORMULAE

1. **If $v = \int a\, dt$ derive $v = u + at$**

   $v = \int a\, dt$
   $v = at + c$
   When $t = 0, v = u$
   $\therefore c = u$
   Hence $v = u + at$

2. **If $s = \int v\, dt$ derive $s = ut + \frac{1}{2}at^2$**

   $s = \int v\, dt$
   $s = \int (u + at)\, dt$
   $s = ut + \frac{1}{2}at^2 + c$
   When $t = 0, s = 0$
   $\therefore c = 0$
   Hence $s = ut + \frac{1}{2}at^2$

3. **Given that $v = u + at$ and $s = ut + \frac{1}{2}at^2$ derive $v^2 = u^2 + 2as$**

Squaring both sides of $v = u + at$
$v^2 = (u + at)^2$
$v^2 = u^2 + 2uat + a^2t^2$
$= u^2 + 2a\left(ut + \frac{1}{2}at^2\right)$
$\therefore v^2 = u^2 + 2as$

4. **Looking at the velocity time graph, derive the distance travelled $s = \frac{(u+v)t}{2}$**

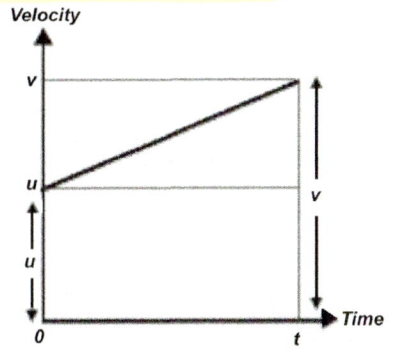

Area under the velocity-time graph is the distance travelled
$s = $ area of trapezium $= \frac{(u+v)t}{2}$

ANSWERS –MOMENTS

MOMENTS
1. The moment of a force acting on a body is the measure of the turning effect of the force on the body.

MECHANICS MEMORY JOGGER

2. The moment of a force (Nm) = force (N) × perpendicular distance from the pivot (m).
   Moment of F about $P = Fx\ Nm$ clockwise.

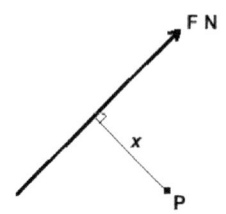

3. The unit of a moment of a force is a Newton-metre (Nm).
4. If a rod is horizontal and in equilibrium then the sum of clockwise moments = the sum of anticlockwise moments.
5. The vertical resolution of forces when a uniform rod of weight W is hanging supported by rope or cables is given by:

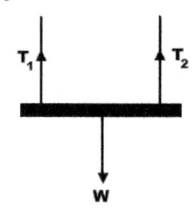

   Equation of motion of uniform rod: $R(\uparrow)$
   $T_1 + T_2 - W = 0$

6. The vertical resolution of forces when a uniform rod of weight W is resting on supports is given by:

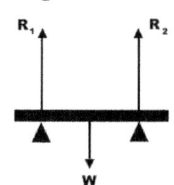

   Equation of motion of uniform rod: $R(\uparrow)$
   $R_1 + R_2 - W = 0$

CENTRE OF MASS

1. The centre of mass of a uniform rod is assumed to be the midpoint of the rod.
2. If the rod is non-uniform, we assume that the centre of mass is not the midpoint of the rod.
3. If a rod is on the point of tilting about one of its supports, the reaction at the other support will be zero.

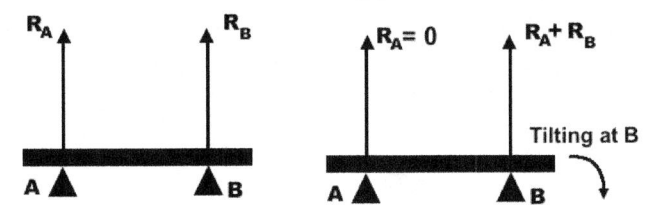

ANSWERS – FORCES AND FRICTION

RESOLVING FORCES

1. The horizontal and vertical components of a force F acting at an angle θ to the horizontal are given by $F\cos\theta$ and $F\sin\theta$ respectively.

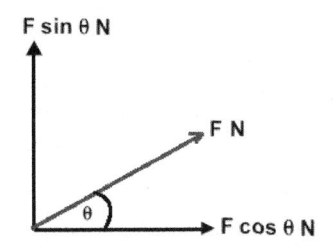

2. The forces acting on an object on a rough plane inclined at an angle θ are resolved parallel and perpendicular to the plane.

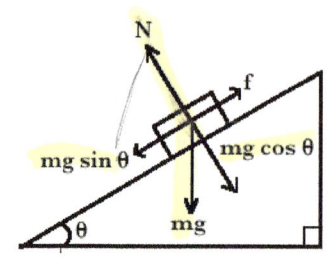

Weight $= mg$
Normal Reaction $N = mg \cos \theta$
Frictional force $f = mg \sin \theta$

FRICTION

1. Friction is a force that opposes the motion between two rough surfaces.
2. The Coefficient of Friction μ for two surfaces in contact is a measurement of the roughness. The rougher the surfaces the larger the μ. Smooth surfaces have $\mu = 0$.
3. The maximum or limiting value of the friction F_{MAX} between two surfaces is given by $F_{MAX} = \mu R$ where μ is the coefficient of friction and R is the normal reaction between the two surfaces.
4. Friction acts in a direction opposite to the direction of motion.

5. Force diagram showing the frictional force.

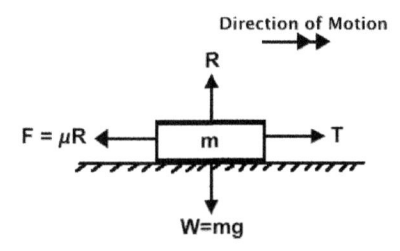

SLOPING PLANES

1. The forces acting on an object on a rough plane inclined at an angle θ are resolved parallel and perpendicular to the plane.

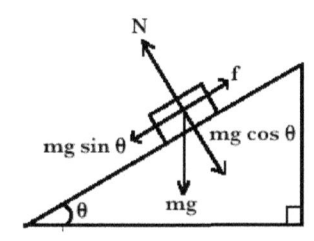

   Weight $= mg$
   Normal Reaction $N = mg \cos \theta$
   Frictional force $f = mg \sin \theta$

2. If the angle of inclination of the slope is given by $\tan \theta = \frac{3}{4}$,
   $$\sin \theta = \frac{3}{\sqrt{3^2 + 4^2}} = \frac{3}{5} \text{ or } \sin\left(\tan^{-1} \frac{3}{4}\right)$$
   $$\cos \theta = \frac{4}{\sqrt{3^2 + 4^2}} = \frac{4}{5} \text{ or } \cos\left(\tan^{-1} \frac{3}{4}\right)$$

3. An object will start to slip down a slope if the component of the weight $mg \sin \theta$ is greater than the friction.

MECHANICS MEMORY JOGGER

ANSWERS – PROJECTILES

VERTICAL PROJECTION
1. For objects projected upwards, $g = -9.8 \, m \, s^{-2}$
2. The total time that an object is in motion from the time it is projected upwards to the time it hits the ground is called **the time of flight**.
3. The initial speed is called **the speed of projection**.
4. The displacement of an object that is projected and returns to its original position is taken as zero ($s = 0$).

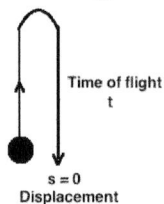

5. If the original projected position is $X \, m$ above the returning position, the displacement is taken as $-X$ ($s = -X$).

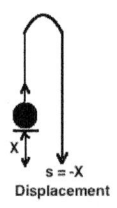

6. If the time of flight is required for the object's motion above a height h, the displacement is taken as h ($s = +h$ or $s = +(h - X)$).

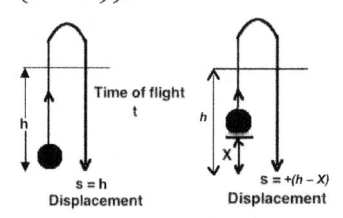

ANSWERS

HORIZONTAL PROJECTION
1. The motion of a particle projected horizontally has only one force $g = +9.8\ ms^{-2}$ acting on it.
2. Forces in a horizontal projection are resolved horizontally $R(\rightarrow)$ and vertically downwards $R(\downarrow)$.

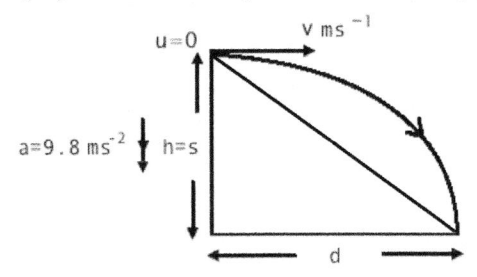

The vertical movement of the projectile is given by
$R(\downarrow): h = \frac{1}{2}at^2$

The horizontal movement of the projectile is given by
$R(\rightarrow): d = vt$

3. The distance of a particle from its point of projection = $\sqrt{h^2 + d^2}$.

PROJECTION AT AN ANGLE
1. If a particle is projected at an angle θ to the horizontal, the initial velocity U is resolved into two components:
$U_x = U \cos\theta$ and $U_y = U \sin\theta$.

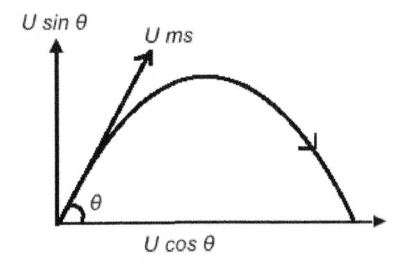

MECHANICS MEMORY JOGGER

2.

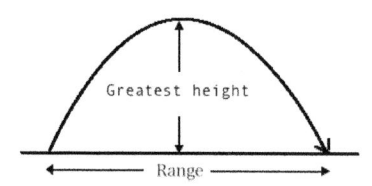

The range is the distance between the point of projection and the point of impact on the horizontal plane.
3. The time taken for the particle to strike the horizontal plane is called the time of flight.
4. The projectile reaches its greatest height when its vertical velocity equals 0.

PROJECTILE MOTION FORMULAE

1. Derive the time of flight.
   Horizontal velocity $U_x = U \cos \theta$
   Vertical velocity $U_y = U \sin \theta$
   Considering upward vertical motion:
   $R(\uparrow)$
   $u = U \sin \theta, s = 0, a = -g, t = T$
   $s = ut + \frac{1}{2}at^2$
   $(U \sin \theta)T - \frac{gT^2}{2} = 0$
   $T\left(U \sin \theta - \frac{gT}{2}\right) = 0$
   $\therefore T = 0 \text{ or } T = \frac{2U \sin \theta}{g}$
   $\therefore$ Time of flight $T = \frac{2U \sin \theta}{g}$

2. Derive the horizontal range.
   Considering horizontal motion:
   $R(\rightarrow)$
   $Speed = \frac{distance}{time}$
   $Range = Ut$
   Horizontal velocity $U_x = U \cos \theta$

$$Range = (U \cos \theta)t$$
$$= U \cos \theta \times \frac{2U \sin \theta}{g}$$
$$= \frac{U^2}{g}(2 \sin \theta \cos \theta)$$
$$\therefore Range = \frac{U^2 \sin 2\theta}{g}$$

3. Derive the greatest height reached.
$$v^2 = u^2 + 2as$$
$$u = U \sin \theta, v = 0, s = h, a = -g$$
$$0 = U^2 \sin^2 \theta - 2gh$$
$$\therefore h = \frac{U^2 \sin^2 \theta}{2g}$$

4. Derive the equation of the trajectory.
$R(\uparrow)$
$$s = ut + \frac{1}{2}at^2$$
$$y = (U \sin \theta)t - \frac{1}{2}gt^2 \quad_____(1)$$
$R(\rightarrow)$
$$x = (U \cos \theta)t$$
Substituting $t = \frac{x}{U \cos \theta}$ in (1) we have
$$y = (U \sin \theta)\frac{x}{U \cos \theta} - \frac{1}{2}g\left(\frac{x}{U \cos \theta}\right)^2$$
$$y = x \tan \theta - \frac{gx^2}{2U^2 \cos^2 \theta}$$
$$y = x \tan \theta - gx^2 \frac{(\sec^2 \theta)}{2U^2}$$
$$\therefore y = x \tan \theta - gx^2 \frac{(1+\tan^2 \theta)}{2U^2}$$ is the equation of the trajectory.

ANSWERS – APPLICATIONS OF FORCES

STATIC PARTICLES
1. A static particle is said to be in equilibrium if the resultant of all the forces acting on it is zero.
2. The body is said to be in limiting equilibrium (on the point of moving) if frictional force $F_{MAX} = \mu R$

MECHANICS MEMORY JOGGER   34

3. The body is said to be at rest and in equilibrium if frictional force $F < \mu R$

STATIC RIGID BODIES

1. Force diagram for a ladder leaning against a smooth wall with one end resting on rough, horizontal ground:

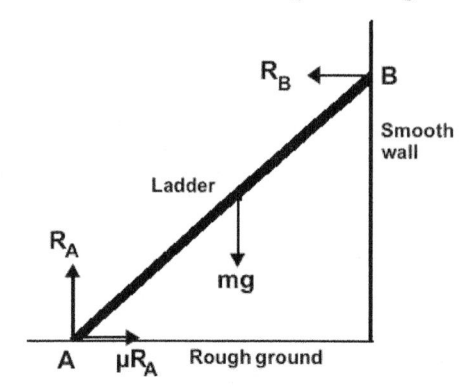

Resolving vertically, $R(\uparrow)$
$R_A = mg$
Resolving horizontally, $R(\rightarrow)$
$\mu R_A = R_B$

2. Force diagram for a uniform ladder leaning against a smooth wall with one end resting on rough, horizontal ground in limiting equilibrium (at the point of slipping):

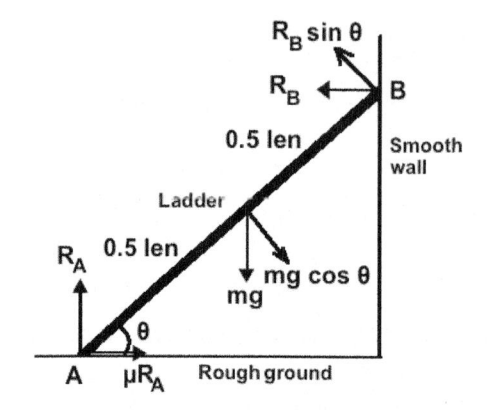

ANSWERS

The ladder is in limiting equilibrium when it is at the point of slipping or frictional force $\mu R_A = R_B$.
Length of ladder $= len$
Weight of ladder $= mg$
Angle made with the horizontal $= \theta$
Resolving vertically, $R(\uparrow)$
$R_A = mg$
Resolving horizontally, $R(\rightarrow)$
$\mu R_A = R_B$
Taking moments about the base of the ladder:
$(0.5\, len) \times mg \cos \theta = (len) \times R_B \sin \theta$

MECHANICS MEMORY JOGGER

3. Force diagram for a uniform ladder leaning against a rough wall with one end resting on rough, horizontal ground in limiting equilibrium (at the point of slipping):

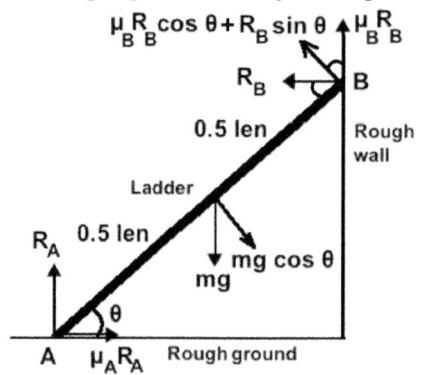

$\mu_B R_B(\rightarrow)$ and $\mu_B R_B(\uparrow)$ are the frictional forces preventing the ladder from slipping.
R_A and R_B are the normal reactions at A and B.
The ladder is in limiting equilibrium when it is at the point of slipping or frictional force $\mu_A R_A = R_B$.

Length of ladder = len
Weight of ladder = mg
Angle made with the horizontal = θ
Resolving vertically, $R(\uparrow)$
$R_A + \mu_B R_B = mg$
Resolving horizontally, $R(\rightarrow)$
$\mu_A R_A = R_B$
Taking moments about the base of the ladder:
$(0.5 \, len) \times mg \cos \theta = (len) \times (R_B \sin \theta + \mu_B R_B \cos \theta)$

ANSWERS – VECTORS IN KINEMATICS

VECTORS IN KINEMATICS
1. A position vector indicates the position of the vector relative to the origin.
If $r = ai + bj$ is a position vector, then the distance of r from the origin is its magnitude $= |r| = \sqrt{a^2 + b^2}$.

ANSWERS

2. A velocity vector describes the direction the object is moving and its speed.
   Speed is the magnitude of a velocity vector.
   If velocity vector $v = pi + qj$, then speed $= |v| = \sqrt{p^2 + q^2}$.

3. Given that a particle starts from a point with position vector r_0 and moves with a constant velocity v. At time t its displacement is vt and its position vector r is given by $r = r_0 + vt$.

VECTORS AND BEARINGS
1. i is the unit vector due east and j is the unit vector due north.
2. To show that two particles collide equate the i and j components of their position vectors at time t.

CALCULUS WITH VECTORS
1. The relationship between displacement, velocity and acceleration to solve problems in 2D vector kinematics is described as follows:

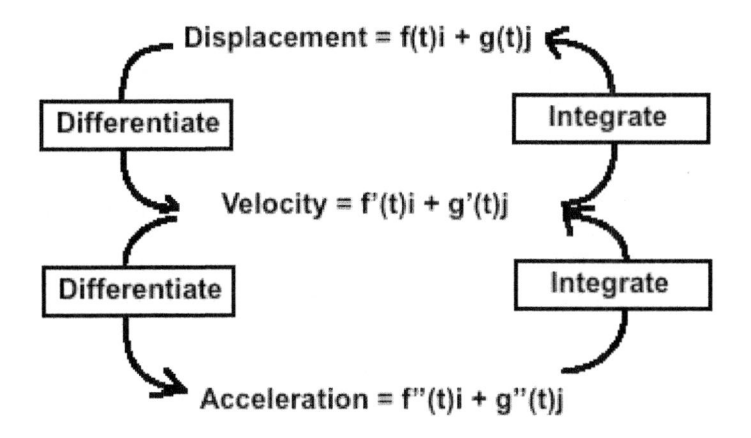

Displacement = f(t)i + g(t)j

Differentiate → Velocity = f'(t)i + g'(t)j ← Integrate

Differentiate → Acceleration = f''(t)i + g''(t)j ← Integrate

2. For vector integration, the constant of integration would be a vector.

ABOUT THE AUTHOR

Shobha Natarajan holds an MSc in Mathematics from Bangalore University and teaches Mathematics to students at A Level, IB and GCSE in the Medway towns of Kent. The Maths Clinic was established in 2011 to publish revision guides in Mathematics in print and Kindle e-book formats. Shobha is a software professional with over 25 years' experience in embedded software development.

Printed in Great Britain
by Amazon